OnBoard ACADEMICS

Fractions

© 2015 OnBoard Academics, Inc
Portsmouth, NH
800-596-3175
www.onboardacademics.com
ISBN: 978-1-63096-068-1

OnBoard Academic's books are specifically designed to be used as printed workbooks or as on-screen instruction. Each page offers focused exercises and students quickly master topics with enough proficiency to move on to the next level.

OnBoard Academic's lessons are used in over 25,000 classrooms to rave reviews. Our lessons are aligned to the most recent governmental standards and are updated from time to time as standards change. Correlation documents are located on our website. Our lessons are created, edited and evaluated by educators to ensure top quality and real life success.

Interactive lessons for digital whiteboards, mobile devices, and PCs are available at www.onboardacademics.com. These interactive lessons make great additions to our books.

You can always reach us at customerservice@onboardacademics.com.

Fractions

Key Vocabulary

half

third

fourth

sixth

twelfth

     www.onboardacademics.com

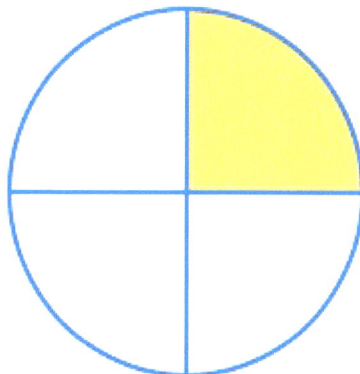

How many parts does this circle have?

How many parts are shaded yellow?

What fraction is shaded yellow?   ——————   one fourth

Fractions

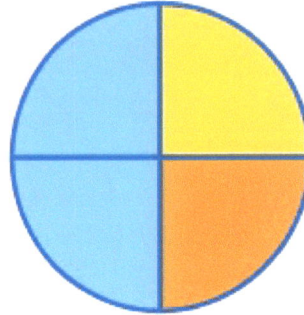

How many parts are shaded blue?

What fraction is shaded blue? ——— two fourths

What fraction is shaded orange? ——— one fourth

www.onboardacademics.com

Here you see that the circle has two fourths or 2/4 shaded blue.

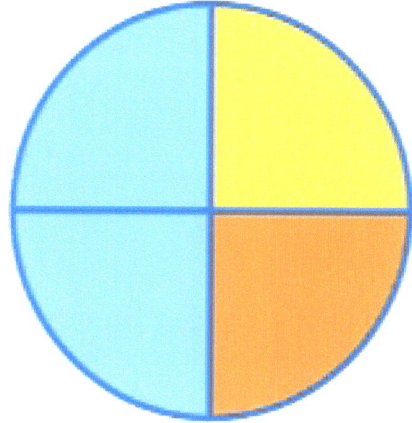

What's another way of writing 2/4?_____

Use the illustration below as a hint.

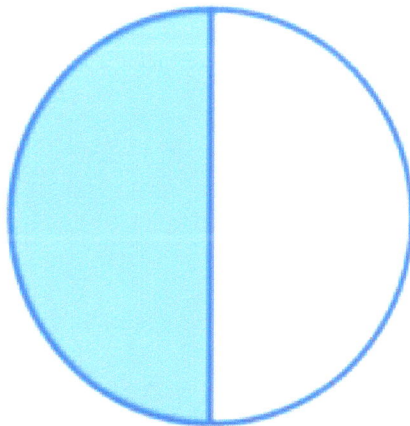

Complete this table.

Is there another way of writing 4/4? _____

Color	# Parts	Fraction
	1	$\frac{1}{4}$
	2	$\frac{2}{4}$
	1	$\frac{1}{4}$
Total		$\frac{}{4}$

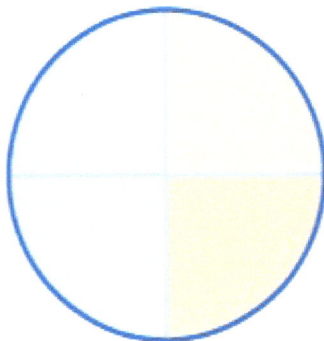

_____

www.onboardacademics.com

What fraction of each shape is shaded?
Fill in your answers.

 ____

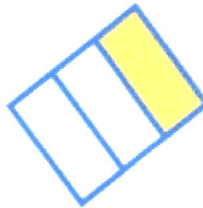 ____

 ____

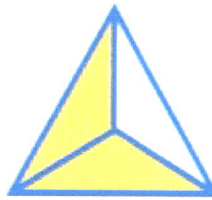 ____

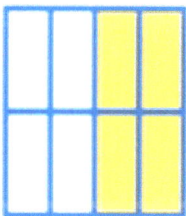 ____

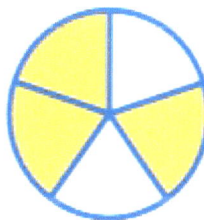 ____

www.onboardacademics.com

What fraction of each shape is NOT shaded?

$$\frac{1}{2} \quad \frac{}{2}$$

$$\frac{1}{3} \quad \frac{}{3}$$

$$\frac{2}{3} \quad \frac{}{3}$$

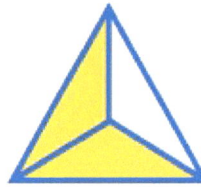

$$\frac{2}{3} \quad \frac{}{3}$$

$$\frac{4}{8} \quad \frac{}{8}$$

$$\frac{3}{5} \quad \frac{}{5}$$

Shade the figure to represent the fraction.

$$\frac{2}{4}$$

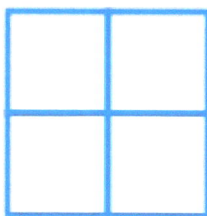

$$\frac{4}{8}$$

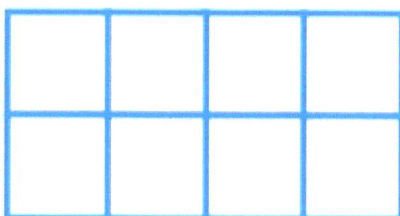

$$\frac{1}{2}$$

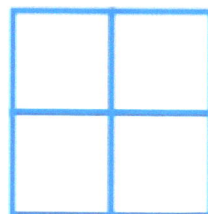

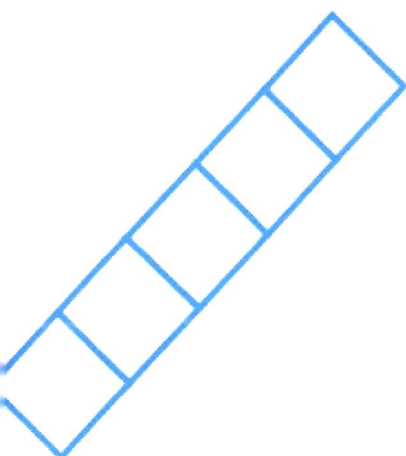

$$\frac{3}{5}$$

$$\frac{1}{3}$$

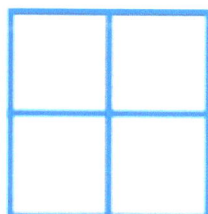

$$\frac{4}{4}$$

www.onboardacademics.com

Red

Green

Yellow

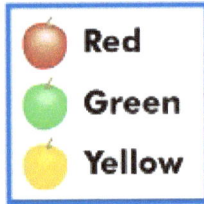

How many apples are there?

What fraction are red? _____

What fraction are green? _____

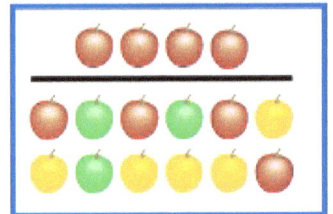

Name_____

Fractions Quiz

Circle or fill in the correct answer.

1 **What fraction of this triangle is shaded?**
Ⓐ $\frac{1}{2}$ Ⓑ $\frac{1}{3}$ Ⓒ $\frac{1}{4}$ Ⓓ $\frac{1}{6}$

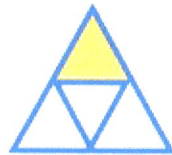

2 **What fraction of this rectangle is shaded?**
Ⓐ $\frac{3}{4}$ Ⓑ $\frac{1}{3}$ Ⓒ $\frac{2}{3}$ Ⓓ $\frac{3}{3}$

3 **What fraction of these apples have been bitten?**

4 **What fraction of this figure has *not* been shaded?**

www.onboardacademics.com

Equivalent Fractions

Key Vocabulary

numerator

denominator

equivalent fractions

Who ate the most pizza?

"I ate two slices of my pizza."

"I wasn't that hungry. I just ate one slice of my pizza."

They both ate the same amount. $\frac{2}{4} = \frac{1}{2}$ $\frac{2}{4}$ and $\frac{1}{2}$ are *equivalent fractions*.

Which figures have one half 1/2 shaded?
Place a check mark int the box if 1/2 shaded and an X if not.
Circle the equivalent fractions.

Finding Equivalent Fractions
Discover equivalent fractions below. What is the first fraction multiplied by in example 2 and 3 to make the equivalent fraction?

> To find an equivalent fraction, multiply (or divide) the numerator and the denominator by the same number.

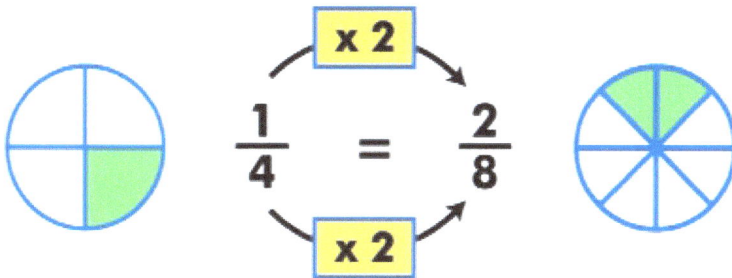

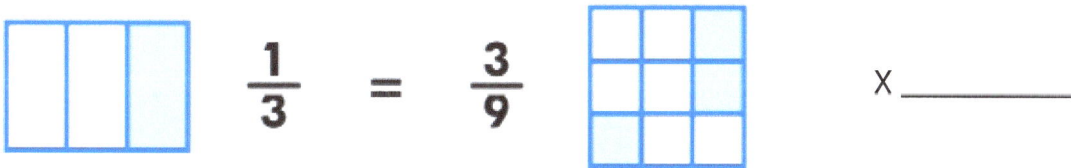

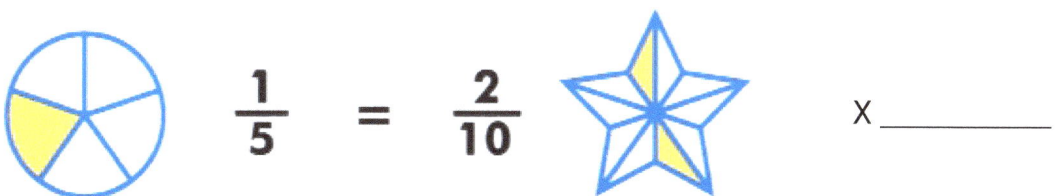

x 2

$$\frac{1}{4} = \frac{2}{8}$$

x 2

$$\frac{1}{3} = \frac{3}{9}$$ X _____

$$\frac{1}{5} = \frac{2}{10}$$ X _____

Find Equivalent Fractions

Fill in the missing boxes using the chart below.

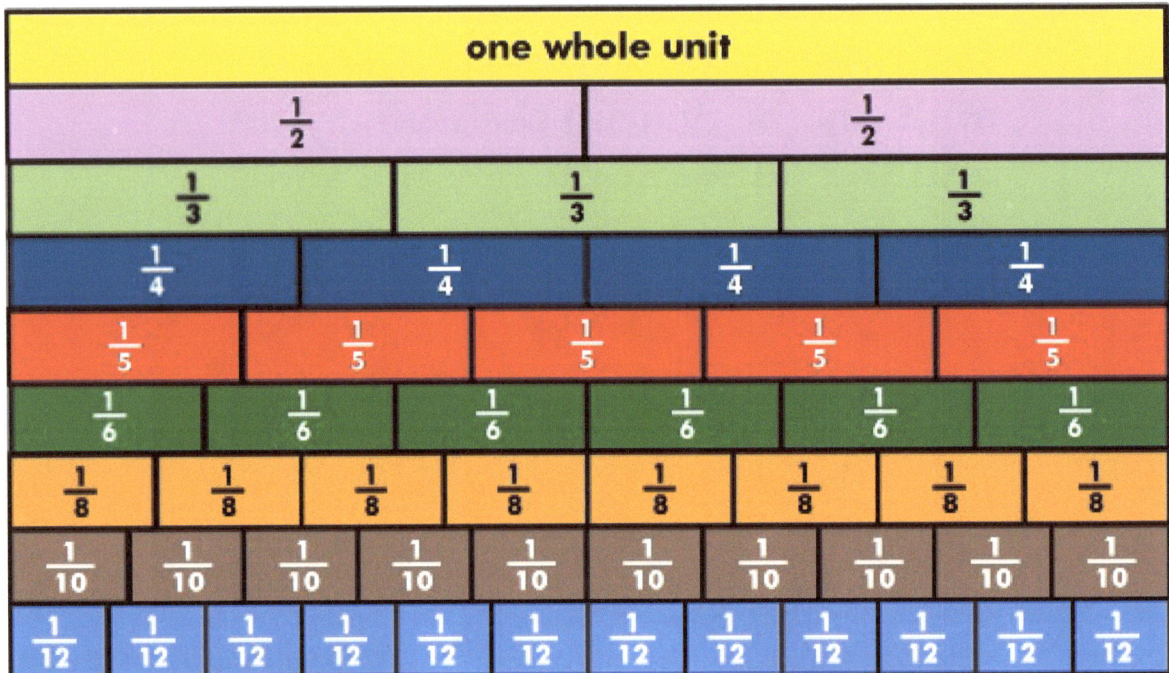

Find Equivalent Fractions

Fill in the empty boxes using the chart below.

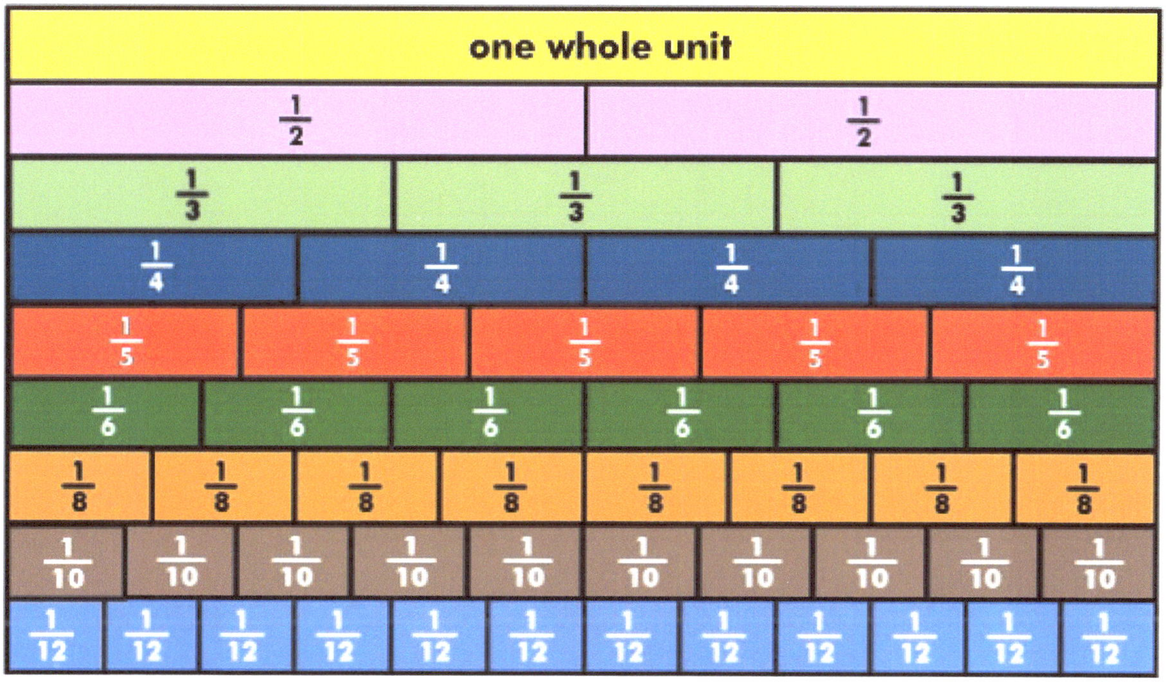

Find Equivalent Fractions

Draw the number cards on the number line. One example has been completed for you.

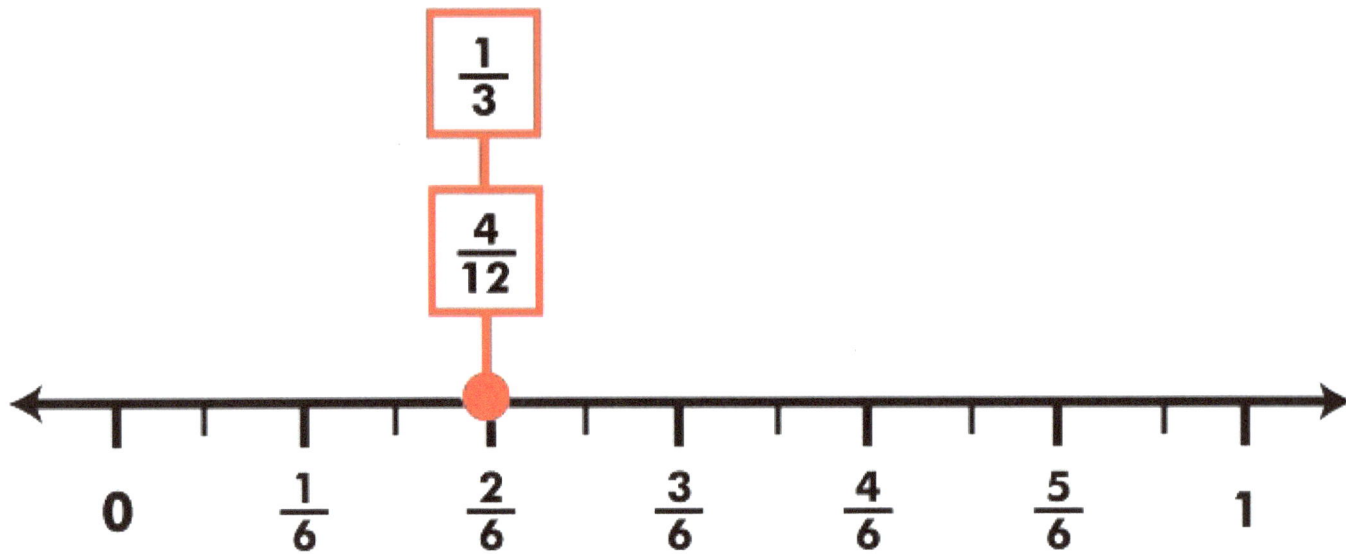

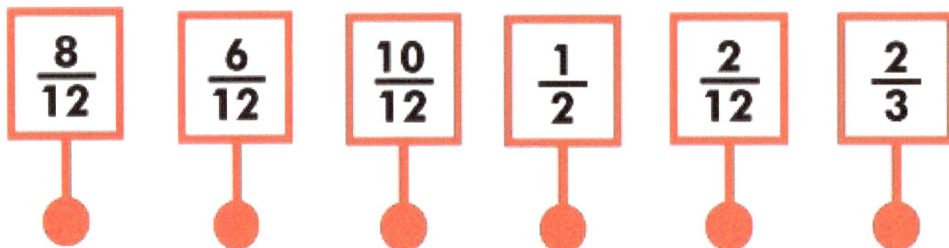

$$\frac{1}{3}$$

$$\frac{4}{12}$$

0    $\frac{1}{6}$ $\frac{2}{6}$ $\frac{3}{6}$ $\frac{4}{6}$ $\frac{5}{6}$ 1

$$\frac{8}{12} \qquad \frac{6}{12} \qquad \frac{10}{12} \qquad \frac{1}{2} \qquad \frac{2}{12} \qquad \frac{2}{3}$$

Name_____

Equivalent Fractions Quiz

1 These two fractions are equivalent, true or false?

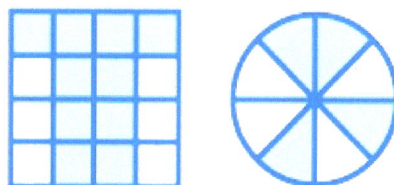

2 Which fraction is the odd one out?

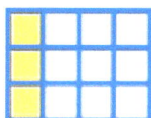

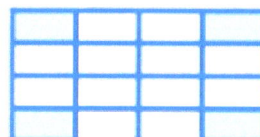

A   B   C   D

3 What is the missing numerator? $\dfrac{5}{10} = \dfrac{?}{40}$

4 What is the missing denominator? $\dfrac{3}{8} = \dfrac{9}{?}$

Add & Subtract Fractions

with Like Denominators

Key Vocabulary

Numerator

Denominator

Equivalent Fraction

How much of the pizza has been eaten?

Add up the total number of pieces eaten and write as a fraction.

How much piece is left? Express your answer in a fraction.

I ate $\dfrac{3}{8}$ of the pizza

I ate $\dfrac{1}{8}$ of the pizza

I ate $\dfrac{3}{8}$ of the pizza

Adding fractions with like denominators.

Color in the columns in the last model to solve the problem.

Practice adding fractions with like denominators.

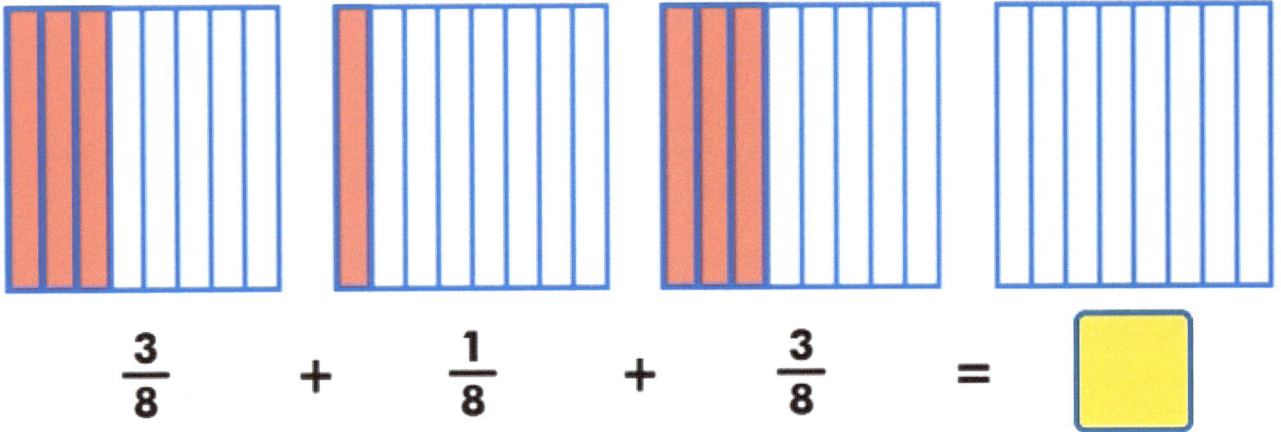

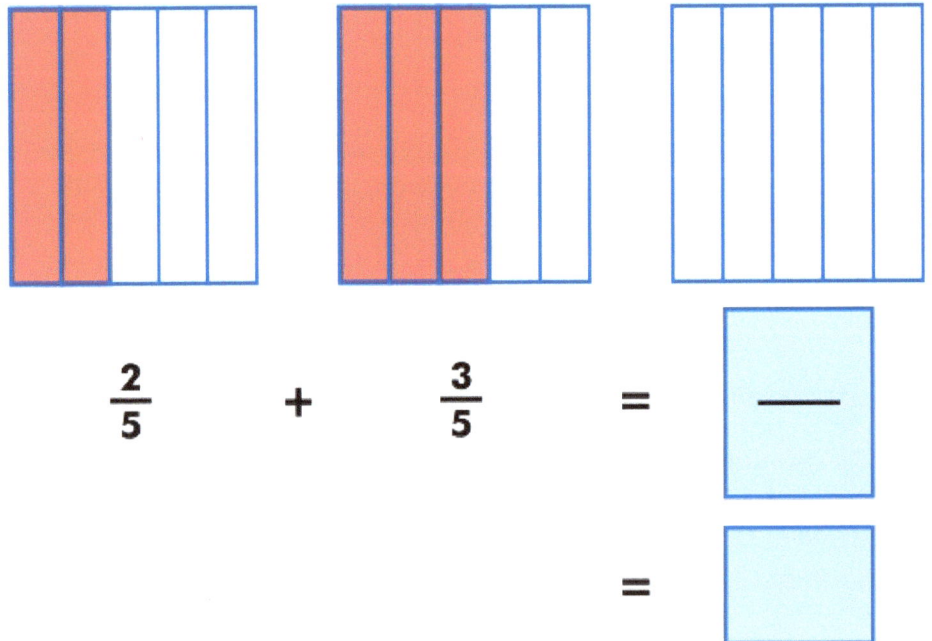

$$\frac{3}{8} \quad + \quad \frac{1}{8} \quad + \quad \frac{3}{8} \quad =$$

$$\frac{2}{5} \quad + \quad \frac{3}{5} \quad = \quad \underline{\quad\quad}$$

$$=$$

Writing answers in the simplest form.

Shade the final model by adding the 10^{ths}. Two 10^{ths} fit into each block in the final model. Find the answer in 10^{ths} and then find a simpler answer using the final model as a clue.

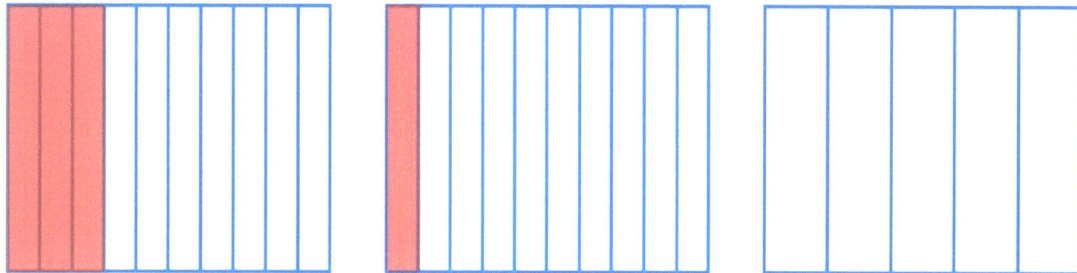

$$\frac{3}{10} \quad + \quad \frac{1}{10} \quad = \quad \boxed{\underline{\quad}}$$

$$= \quad \boxed{\underline{\quad}}$$

Subtracting Fractions with Like Denominators

Use the model to complete the problem. After you complete the problem, express the fraction in a simpler term.

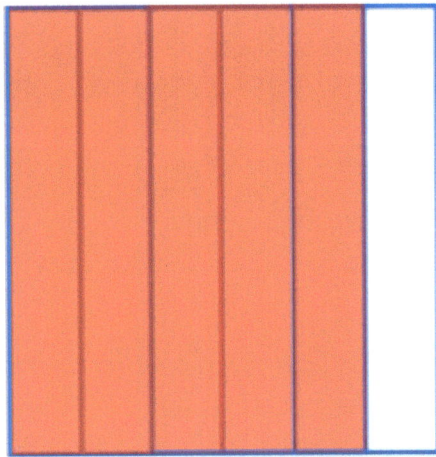

$$\frac{5}{6} - \frac{1}{6} = \boxed{}$$

$$= \boxed{}$$

Practice adding and subtracting fractions with like denominators.

After you complete the problem, express the answer in a simpler fraction.

① $\dfrac{1}{6}$ + $\dfrac{5}{6}$ = ☐ = ☐

② $\dfrac{5}{9}$ − $\dfrac{2}{9}$ = ☐ = ☐

③ $\dfrac{9}{10}$ − $\dfrac{4}{10}$ = ☐ = ☐

④ $\dfrac{19}{21}$ − $\dfrac{5}{21}$ = ☐ = ☐

Solve this fraction word problems.

$\dfrac{3}{12}$ of Mrs. Jones' class were absent with flu. $\dfrac{5}{12}$ of her class were on a field trip and $\dfrac{1}{12}$ of her class were at an award ceremony. Mrs. Jones' sister, the district superintendent, was presenting the awards. What fraction of Mrs. Jones' class were not in school?

What fraction of Mrs. Jones' class were in school?

Name_____

Add & Subtract Fractions with Like Denominators Quiz
Circle or fill in the correct answer.

1 True or false? $\frac{1}{5} + \frac{3}{5} = \frac{4}{10}$

2 $\frac{4}{6} + ? = 1$

   A $\frac{1}{2}$ **B** $\frac{6}{2}$ **C** $\frac{4}{6}$ **D** $\frac{1}{3}$

3 $\frac{10}{11} - ? = \frac{4}{11}$

4 Sarah spent $\frac{5}{9}$ of her money on clothes, $\frac{1}{9}$ on the movies and another $\frac{1}{9}$ on ice cream. Her movie ticket cost $12. What fraction of her money did she have left?

www.ingramcontent.com/pod-product-compliance
Lightning Source LLC
Chambersburg PA
CBHW052045190326
41520CB00002BA/199